CW01370936

DATE DUE			
JUN 12 1967			
MAY 16 1967 M P			
APR 23 '82 O			
MY 7			
NOV 08 1994			
NOV 3 1997			

Cornell University Library
Z8715 .S94
Ptolemy's geography: a brief account of

3 1924 029 642 414

olin

PTOLEMY'S GEOGRAPHY

PTOLEMY'S GEOGRAPHY
A BRIEF ACCOUNT OF ALL THE
PRINTED EDITIONS
DOWN TO 1730
WITH NOTES ON SOME IMPORTANT VARIATIONS
OBSERVED IN THAT OF ULM 1482
INCLUDING THE RECENT DISCOVERY OF
THE EARLIEST PRINTED MAP OF THE WORLD
YET KNOWN ON MODERN GEOGRAPHICAL CONCEPTIONS
IN WHICH SOME ATTEMPT WAS MADE
TO DEPART FROM ANCIENT
TRADITIONS

By HENRY N. STEVENS, F.R.G.S.

SECOND EDITION

LONDON
HENRY STEVENS, SON AND STILES
39 GREAT RUSSELL STREET
OVER AGAINST THE SOUTH-WEST CORNER OF
THE BRITISH MUSEUM
1908
JF

CHISWICK PRESS: CHARLES WHITTINGHAM AND CO.
TOOKS COURT, CHANCERY LANE, LONDON

PREFATORY NOTE

The first edition of this little Essay was privately printed in an *édition de luxe* of only Twenty copies, as a descriptive handbook to the HENRY STEVENS COLLECTION of the various editions of Ptolemy's Geography, now preserved in the Library of Mr. Edward E. Ayer of Chicago. Every one of these twenty copies included a set of the original impressions of the Sixty Bookplates, each of which had been separately and specially prepared for insertion in the particular volume of the Collection to which it appertained.

I had no intention of publishing these Notes until several friends urged me to do so, on the ground that the information as to the curious variations in the Ulm Ptolemy of 1482 and the recently discovered early World Map, was of sufficient interest and importance to students of early typography, bibliography and cartography to be worthy of being placed on record.

Having agreed to print only Twenty copies of the separate Bookplates, it has not been possible to reprint them here in their original form. I have therefore contented myself with reproducing the first as a specimen, and adding merely the descriptive text as it appeared on the remainder, so as to form as it were a brief chronological List of the various editions of Ptolemy's Geography with the salient points of each.

HENRY N. STEVENS.

39, Great Russell Street,
    London, W.C.
        20 *October*, 1908.

THE HENRY STEVENS PTOLEMY COLLECTION

This magnificent Collection was commenced by my late father about the year 1848, shortly after he first came to London from Vermont on bibliographical investigations intent. His general design was to illustrate cartographically the progress of modern geographical discovery and knowledge, by a series of the editions of Ptolemy's Geography (as many as he could obtain) from pre-Columbian down to comparatively recent times. More particularly was he desirous of tracing the gradual cartographical "evolution of America from the sea of darkness." The Collection was originally intended for his own use in his general researches in this direction; and by the time of his death in 1886 he had acquired perhaps thirty different editions. At the same time, while collecting for himself, he largely built up the splendid sets of Ptolemy's Geography in the Libraries of Mr. James Lenox of New York, Mr. John Carter Brown of Providence, and the British Museum.

In dealing with my father's Library after his death, as his literary executor, it seemed a pity to disperse a series of books which had been gotten together with such patient care; so I formed the design of endeavouring to complete it as a Henry Stevens Memorial Collection. During the next nine or ten years some sixteen more editions were added, and as the volumes were in all sorts of bindings, good, bad, and indifferent, it was thought that the Collection would be much improved and be more likely to be kept intact if the whole was in uniform binding. I accordingly conferred with my old friend Mr. Pratt, who, by the way, with his father before him, has bound for me and my father for upwards of fifty years. After much consultation, a fairly plain and substantial binding rather than a showy one was decided on; olive green morocco of the finest possible quality, elegantly tooled in blind and gold. Owing to the great difference in size and shape of the various editions between the smallest octavo of Cologne 1540 and the largest folio of Amsterdam 1730, it became necessary to design a pattern which would expand or contract to suit the various shapes and sizes of octavo, quarto, and folio; and the ornamenting tools had to be cut in several sizes to correspond.

By the time the binding of the forty-six volumes had been completed in the spring of 1898, the expense was getting altogether beyond my means, especially so as there were still some very important and expensive editions yet to be acquired if the idea was to be carried to a successful conclusion. Accordingly I began to look round for a patron who would be willing to co-operate with me, not only by purchasing the Collection as it stood, but who would also undertake to keep it intact and add to it when opportunity offered. Mr. Edward E. Ayer of Chicago happening to be in London about this time, the matter was mentioned to him, and very speedily the business was concluded to our mutual satisfaction, especially so to mine, for he at once very cordially fell in with my desire to perpetuate the Henry Stevens Collection and to add to it. Since it was transferred to him in 1898, Mr. Ayer has nobly redeemed his promises, with the result that fourteen most valuable editions or variations have been added, bringing the total number of volumes up to sixty. At the present time only one or two editions of importance are still wanting.

Every volume contains a specially designed Memorial Bookplate, each one recording briefly the salient points of the particular edition to which it appertains. The border of this Bookplate, in the Holbein style, was copied and adapted from one of the map borders of the Basle Edition of 1540. In the preparation of the short abstracts on each Bookplate, free use has been made of Mr. Wilberforce Eames's bibliography of Ptolemy's Geography,[1] and I have great pleasure in taking this opportunity of expressing my obligation to him and tendering my thanks. The reader is cordially referred to Mr. Eames's valuable work for the further details of each edition, which the limited capacity of the Bookplates and the following Essay necessarily excluded. A reproduction of the first Bookplate is appended hereto

[1] Eames (Wilberforce) *A List of the Editions of Ptolemy's Geography, 1475-1730, New York,* 1886, also included in Sabin's Dictionary.

as a specimen, followed by the descriptive text as it appeared in similar style on the remainder. This, it is hoped, will form a handy chronological reference list to the various editions of Ptolemy's geography. From this list it will be seen how marvellously full the Collection is, and it is to be hoped the very few remaining "desiderata" may yet be acquired.

The following Essay traces briefly in narrative form the origin and history of the various editions of Ptolemy's Geography, dividing them into sections or groups and particularising the most important. It amplifies the information given on the Bookplates, and shows how the maps gradually increased in number from twenty-seven to sixty-nine. It is to be understood that all the editions hereinafter referred to are represented in the Collection unless otherwise stated. Of special interest will be found the new information relating to recently discovered variations in the map of the world, and in the text of the Ulm Edition of 1482.

PTOLEMY'S GEOGRAPHY

"LET us go back for a moment and survey the little old world as it appeared about the middle of the fifteenth century. According to Ptolemy, the best recognized authority, whose geography had stood the test of thirteen hundred years, the then known world was a strip of some seventy degrees wide, mostly north of the equator, with Cadiz on the west, and farthest India or Cathay on the east, lying between the frozen and burning zones, both impassable by man."

HENRY STEVENS, in *Historical and Geographical Notes*, 1869.

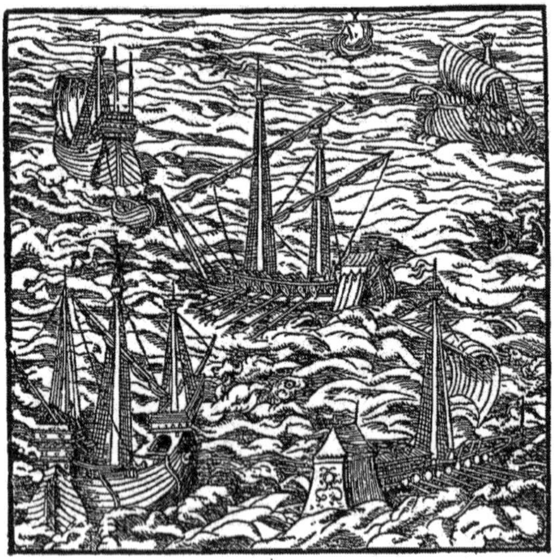

PTOLEMY'S GEOGRAPHY

Prior to the year 1475, when the first printed edition with a definite date appeared at Vicenza (without maps), the geography of Ptolemy in manuscript form had for centuries been regarded as the standard work on the subject of geography in general. Several of these early manuscripts of almost priceless value are still preserved, and some of them, or others of a similar nature, no doubt formed the bases for the maps found in the earliest printed editions. Professor Joseph Fischer, in a recent work,[1] gives a long and interesting account of a number of these manuscripts, to which the reader is referred. No sooner had the idea been conceived of printing Ptolemy's text with maps, than four separate and distinct editions appeared almost simultaneously, three in Italy and one in Germany. These four sets of maps comprise four distinct types, differing not only in number but in design,

---

[1] Fischer (Joseph), *The Discoveries of the Norsemen in America with special relation to their early cartographical representation.* London. Henry Stevens, Son & Stiles. 1903. Roy. 8vo, *xxiv + 132 pp. + 10 facsimiles of old maps.*

projection and the character of the engraving. They are found in the Rome edition of 1478 (27 maps); in the metrical version of Berlinghieri, first issued at Florence about 1480 (31); in the Bologna edition of 1462[1] [supposed error for 1472 or 1482] (26); and in the Ulm issue of 1482 (32). In the first three of these the maps are engraved on copper plates, and authorities disagree greatly in their opinions as to their respective priority, owing to the differing character of the engraving and the great uncertainty of the dates of issue of the Florence and Bologna editions.

Whichever of these three editions may be the earliest, suffice it to say that all the maps, except the modern ones in Berlinghieri, were doubtless derived or redrawn from manuscripts then extant, examples of which still survive.[2] These manuscripts probably in their turn were derived from one common prototype lost in antiquity. The maps in the Ulm edition of 1482 are again quite distinct in character, and are engraved on wood instead of on copper. In the case of this Ulm edition it is fortunately possible to identify almost positively the actual manuscript from which it was derived, as described later.[3] These four issues then may be stated to form the printed prototypes of the ancient maps in most of the later printed editions, for as the printed versions came more and more into general use, the old manuscripts dropped more and more into disuse. Each of these four editions had been prepared by different editors (as detailed in the appended list), and no doubt tended largely

---

[1] This edition has yet to be secured for the Collection.
[2] Cf. Fischer, *op. cit.*
[3] *Vide* page 30.

Ptolemy's Geography

to sow the seeds of future geographical research. The demand for knowledge of the progress of modern geographical discovery became so great that during the next 250 years Ptolemy's Geography, in its constantly improved form, still continued to be the standard work on the subject, so much so that upwards of fifty more editions or collateral works appeared before 1730. To gather these all in one chronological sequence has been our aim in forming the Henry Stevens Collection.

Almost every succeeding edition was re-edited, improved and expanded by new text, annotations and commentaries by the most prominent geographical savants of their times, and of various nationalities. The old maps were continually redrawn to suit the shape and size of the new editions, and new maps were added from time to time as the progress of modern discovery and knowledge rendered necessary.

It is interesting to trace rapidly the effect of the four printed prototypes already referred to. But these four must first be subdivided into two important groups—one in which were included only the ancient maps as found in the old manuscripts, and the other in which some attempts were made to show the advance of modern geographical knowledge by the addition of modern maps. In the former of these two groups we have the editions of Bologna, 1462 [1472 or 1482?], and Rome, 1478, while in the latter we have Berlinghieri's metrical version [c. 1480], and the Ulm issue of 1482. To the Ulm volume has usually been assigned the merit of being the first edition in which an attempt had been made to bring Ptolemy up to date by the introduction of modern maps, but that honour undoubtedly belongs to Berlinghieri's work,[1] where we find new maps of Italia, Hispania, Gallia and Palestina. It is also noteworthy as the only edition in which the old Ptolemeian maps are reproduced on the original projection with rectangular and equidistant meridians or parallels as used in the earliest manuscript maps. In the editions of Rome, Bologna and Ulm, the maps are all redrawn on new projections with slanting or curved meridians and parallels.

To follow exactly the derivatives from the maps in the Berlinghieri and Bologna editions is somewhat difficult, but doubtless they exercised some effect on later issues. But the influence of the maps of Rome 1478, and Ulm 1482, is clear and distinct, and each of these editions was the parent of many subsequent ones. The Ulm edition of 1482 shows considerable advance on Berlinghieri's, for it contains five new maps instead of the four mentioned above. The new addition is a most important map of Northern Europe showing Greenland, which fact Scandinavian geographers make much of as proving that the new world was known long before the discoveries of Columbus. The Ulm map of the world is also most noteworthy, for it is the first in which any attempt is made to depart from the old Ptolemeian representation by showing modern discoveries. This map is further remarkable from the fact that it is the first to bear an inscription denoting the engraver's name, "Insculptum est per Johannē Schnitzer de Armszheim." Even this has led to differences of opinion, for some authorities regard Schnitzer as a surname, while others more properly regard

---

[1] Cf. Nordenskiöld (A.E.). *Facsimile Atlas.* Stockholm. 1889. *Folio.* Pp. 12-14.

Ptolemy's Geography

it as the trade description of John of Armszheim. Respecting an earlier state of this map without the engraver's name I have something further to say, and revert to it later[1] in describing more particularly the two different copies of the book in this Collection. The maps in this Ulm edition were the work of the editor Donnus Nicolaus Germanus, who is often erroneously cited as "Donis."[2]

In 1486 there appeared the second edition of Ulm, reprinted there with the same thirty-two maps, but with the addition of a *Registrum* and a new treatise *De Locis*, usually ascribed to the original editor Donnus Nicolaus Germanus. Fischer, however, points out that the *Registrum* was certainly the work of Iohann Reger, whose name appears in the Colophon, and he holds it possible that Reger was also the author of the treatise *De Locis* as well, but in any case Donnus Nicolaus Germanus was not.[3]

It is said by most authorities that both the index and treatise had already been added (in a different and earlier impression than that of 1486) to some copies of the 1482 edition, but I have not as yet succeeded in finding a copy containing them.

Following in chronological sequence we next have the second Rome edition, reprinted there in 1490, with the addition of the index and treatise, copied from the Ulm edition of 1486. The maps, which comprise only the twenty-seven ancient ones, are reprinted without alteration from the same copperplates as used in 1478. In 1507 appeared the third Rome edition revised and re-

---

[1] *Vide* pp. 28-29.   [2] Cf. Fischer, *op. cit.*, p. 72, *etc.*
[3] Fischer, *op. cit.*, pp. 77-78.

edited. The twenty-seven old maps re-appear, again printed from the old plates, but six new modern maps are now added, viz., Northern Europe, Spain, France, Poland etc., Italy, and the Holy Land. In 1508 the remaining stock of the edition of 1507 was re-issued at Rome with a new title-page, and with the addition of a treatise by Beneventanus, in which the New World is first described in any edition of Ptolemy's Geography. This re-issue of 1508 also contained the famous new map of the world by Johann Ruysch which, until the recent discovery in 1901 of the large Waldseemüller map of 1507,[1] was considered to be the earliest printed map showing any part of the New World. Now it can only be described as the first map in any edition of Ptolemy's Geography showing the New World. Our book-plate for the edition of Rome, 1508, having been printed prior to the discovery of the Waldseemüller map, requires modification to this extent.

Some authorities hold that Ruysch's map was issued in some copies of the edition of 1507, in fact the copy in this Collection contains it. But it is now generally supposed that the map was not really issued till 1508, but that parties who had already purchased the edition of 1507 obtained copies subsequently for insertion in their books. Two states of this map are known, one with and one without the words *Plisacus Sinus* off the eastern coast of Asia; but authorities differ as to which is the earlier. Both the copies in this collection contain the words. Personally, I am inclined to

---

[1] Cf. *The oldest map with the name America of the year* 1507, etc. Edited by Prof. Jos. Fischer, S.J., and Prof. Fr. R. v. Wieser. *London.* Henry Stevens, Son & Stiles. 1903. *Folio.*

think the state with the words is the earlier, on account of the curious hatching on the plate which does not contain them. One would expect to find just such an appearance if the words had been hammered out on the copper plate and the resulting blank space roughly shaded over.

In 1511 appeared the first Venice edition, edited by Bernardus Sylvanus. The twenty-seven ancient maps were re-engraved in quite a new guise, and printed in red and black. A new map of the world drawn on a heart-shaped projection was added, and this, until the recent discovery of the Waldseemüller map of 1507, was considered to be the first printed map to contain the delineation of any part of the Northern Continent of the New World. This statement must now be qualified as applying only to any edition of Ptolemy's Geography, and our bookplate for this edition requires modification accordingly. As far as known no further editions were issued in Italy until an entirely new series was commenced in 1548.

Although not actually an edition of Ptolemy's Geography, there appeared at Cracow in 1512 a little geographical treatise by Ioannes de Stobnicza entitled *Introductio in Claudii Ptholomei cosmographiam* which has always created the greatest interest to modern geographers by reason of the curious map of the world which it contains. The book is an excessively rare one, and has so far eluded our search for this Collection. Of the map it is believed only two copies are known, one at Vienna and the other at Munich. Until the recent discovery of the Waldseemüller map of 1507, this Stobnicza map was held to be the earliest in print in which the North and South Continents of the New World were joined. It has always been a puzzle to geographers as to the source from whence Stobnicza derived his ideas, and great credit has been given to him, not only for his portrayal of the New World at that early date, but also for his supposed new method of projection in two hemispheres. We now know that the Polish savant made merely an almost exact copy of the two inset hemispheres at the top of the large Waldseemüller map of 1507.[1] It is not quite certain when Stobnicza's work was first published, as copies seem to vary,[2] but it was certainly reprinted in 1519, a copy of which edition is in this collection. It is uncertain whether the map was re-issued with this second edition, at all events no known copy of the book contains it. A facsimile of the map as found in the 1512 edition has however been added to the copy of the 1519 edition in this Collection for the purpose of reference and comparison.

Ert we come to the Strasburg edition of 1513, which is generally considered to be the most important of all, by reason of the twenty modern maps which it contains. The true history of this most interesting edition has yet to be written. Suffice it to say here that probably no other early geographical work has led to so

---

[1] Cf. The Waldseemüller map facsimiles, also the facsimile of the Stobnicza map in Nordenskiöld's *Facsimile Atlas*, Plate XXXIV.

[2] G. Winsor, *A Bibliography of Ptolemy's Geography*, 1884, p. 10, and Nordenskiöld's *Facsimile Atlas*, p. 68, etc.

much discussion, and to so many differences of opinion amongst modern geographical writers. Most of these opinions can be traced to the curiously different deductions drawn from certain passages in the book, each authority having translated them differently to suit the various theories he was desirous of proving.

But on one point all writers seem agreed, and that is that this edition was actually projected and commenced by Martin Waldseemüller and his *confrères* at St. Dié as early as 1505. Many authorities say that some of the maps were actually engraved at that period. In support of these contentions it may be mentioned that some few years ago I discovered a map of the world[1] which was evidently an earlier impression than the one found in the 1513 Strasburg Ptolemy. This map I believe was prepared by Waldseemüller at St. Dié prior to 1507 for the then projected edition of Ptolemy, and it is doubtless the earliest printed map not only to show the New World discoveries, but also to bear the name America. When Professor Fischer's discovery in 1901 of the large Waldseemüller map of 1507 was first announced, my friend Mr. Basil H. Soulsby, Superintendent of the Map Room, British Museum, immediately set forth my theories in an article in the "Geographical Journal" for February, 1902, to which the reader is referred. Since the publication of the Waldseemüller facsimiles I have devoted much time to a re-study of the whole matter by the light of the newly discovered map, but see no reason for altering my opinion that my map is earlier than the large map of 1507, and consequently is the earliest printed map yet discovered which shows and names America. I can only hope that I may in time find sufficient leisure from business to enable me to complete and pub-

[1] Now in the John Carter Brown Library, Providence.

lish my essay on this most important subject. Whether I be right or wrong I can promise a mass of new information of a most interesting nature.

Reverting to the Strasburg Ptolemy of 1513, its importance is readily to be observed. The work is divided into two parts, the first being confined to the text of Ptolemy with the twenty-seven ancient maps re-drawn and re-engraved on wood somewhat after the style of the Ulm maps of 1482 and 1486; while the second part includes the twenty modern maps specially prepared for this edition. The modern maps include the *Tabula Terre Nove*, one of the earliest maps devoted specially to the New World.

In 1520 appeared the second Strasburg edition, containing the same forty-seven maps printed from the same woodblocks, but several omissions are made in the text.

The third Strasburg edition, somewhat smaller in size, was published in 1522, re-edited by Laurentius Fries, with some additions to the text. The maps were entirely re-drawn and re-engraved on wood on a reduced scale (except Asia V), and three new ones added, bringing the total up to fifty. The map of the world is noteworthy as being the first in any edition of Ptolemy's Geography to bear the name AMERICA.

In 1525 the fourth and last Strasburg edition was issued, newly translated and re-edited. The fifty maps are from the same blocks as in the 1522 edition except that Asia V has now been re-engraved the same size as the others. Although no more editions were published at Strasburg, the woodblocks of the fifty maps were transferred to Lyons and used again in 1535 in the first edition edited by Michael Villanovanus, better known as Servetus. The map blocks were again used in 1541 in the second Servetus edition printed at Vienne in Dauphiné, by Gaspar Trechsel.

Eantime, during the currency of the Strasburg-Lyons-Vienne editions (1513-1541), several versions of Ptolemy's text without maps had appeared. In 1514 Joannes Stuchs printed at Nuremberg a new Latin translation by Joannes Werner, with the addition of several geographical treatises by modern writers. In 1533 the Greek text was published for the first time, having been printed by Hieronymus Froben at Basle in a handsome quarto volume. This Greek text was newly translated into Latin by Joannes Noviomagus, and printed in 8vo form at Cologne in 1540 by Ioannes Ruremundanus.

Ith the year 1540 an entirely fresh series of maps was inaugurated by the appearance at Basle of a new and important edition, re-edited by the celebrated geographer Sebastian Münster, and printed by the famous printer Henricus Petri. The maps, forty-eight in number, were entirely redrawn and re-engraved, and comprise the twenty-seven ancient ones according to Ptolemy, and twenty-one new ones in which the modern discoveries are more correctly laid down than in any of the preceding editions. These maps were probably engraved on wood, but the letterings are apparently printed from metal type.

A second Basle edition appeared in 1542 from the same

press, reprinted without alteration. A third edition was issued by the same printer in 1545 with the addition of six new maps, raising the total to fifty-four. Münster's work was evidently very popular, for a fourth edition became necessary in 1552. This was also printed by Petri at Basle, with the same fifty-four maps as in the third issue, but with some additions to the text.

Münster died in 1552 and it is somewhat remarkable to find, after a lapse of nearly twenty years, that twenty-four of the maps which he had prepared to illustrate Ptolemy's Geography, were printed again from the old blocks to illustrate a Latin edition of Strabo's Geography, published by Petri at Basle in 1571. The famous Greek geographer Strabo flourished a century or more before Ptolemy, and a copy of this 1571 edition of his work has been added to this Collection, not only to show the use made of Münster's Ptolemy maps in illustrating that work, but also to afford an opportunity of comparing the texts of Strabo and Ptolemy.

Between the third and fourth editions of Münster (1545 and 1552) Ptolemy's text had been translated into Italian and published for the first time in that language[1] at Venice in 1548 in a thick small 8vo volume. The maps, sixty in number, comprise twenty-six ancient, and thirty-four modern ones, all elegantly engraved on copper in an entirely new style. They were designed by Jacopo Gastaldo, and are based for the most part on the maps of Münster, but with considerable additions. This edition does not seem to have been

[1] Berlinghieri's work of 1480(?) did not contain the actual text of Ptolemy.

reprinted, probably the size was found to be too small for practical purposes.

New and important Italian edition, edited by Girolamo Ruscelli, was published by Valgrisi at Venice in 4to form in 1561. The maps, based for the most part on those of Gastaldo of 1548, were redrawn and re-engraved on copper on a larger scale, and the number increased to sixty-four, of which twenty-seven were ancient and thirty-seven modern. The second edition of Ruscelli, in Italian, was published by Giordano Ziletti in 1564. It is an edition of considerable rarity, in fact Nordenskiöld confesses to not having seen it.[1] It contains the same sixty-four maps as in the edition of 1561. A third Italian edition of Ruscelli's version, revised and corrected, was issued by Ziletti at Venice in 1574. It includes the same sixty-four maps, with the addition of a new one of Rome. A fourth Italian edition of Ruscelli's version, revised, enlarged, and re-edited by Gioseppe Rosaccio, was published by the heirs of Melchior Sessa at Venice in 1598-1599. The maps include four entirely new ones, increasing the number to sixty-nine, the remaining sixty-five being from the same plates as last used in 1574, but retouched and figures of ships, fishes, etc., introduced.

During the currency of Ruscelli's Italian editions, 1561-1599, several Latin issues had appeared. In 1562 there was published by Valgrisi at Venice a new edition in 4to, revised by Josephus Moletius. This contained the same sixty-four

[1] *Facsimile Atlas,* p. 26, No. 32.

copperplate maps which had already been used in Ruscelli's first Italian edition issued by the same publisher in the previous year. The unsold stock of this Latin edition of 1562 was reissued at Venice in 1564 by Ziletti, with the same sixty-four maps, only the first sheet comprising the title and dedication having been reprinted. This re-issue is excessively rare, and appears to have escaped the notice of Eames, but is mentioned by Nordenskiöld,[1] and is described on our Bookplate.

In 1578 Gerardus Mercator issued at Cologne a folio volume containing twenty-eight maps to illustrate Ptolemy, with an index, but Ptolemy's text was not included. These maps were beautifully engraved on copper, with ornamental borders. They were re-issued again in 1584, this time with Ptolemy's text, the version of Pirckheymer edited by Arnoldus Mylius, by whom the work was dedicated to Abraham Ortelius. A new edition in 4to, with new maps, edited by Giovanni Antonio Magini of Padua, was published at Venice in 1596. The maps, sixty-four in number, were re-engraved by Girolamo Porro on a somewhat smaller scale than those in the Italian and Latin editions of Ruscelli and Moletius, 1561-1574. Magini's second edition, also in 4to, appeared at Cologne in 1597 with the same sixty-four maps. Some copies of this edition have a colophon *Armhemii apud Ioannem Iansonium Bipliopolam* on the verso of the last leaf. Copies of both issues are in the Collection. A third edition of Magini's work in similar form was published at Cologne in 1608 with the same sixty-four maps, and in 1617 a fourth edition was issued, this time with the imprint of Armhem, re-edited by Gaspar Ens, also with the same sixty-four maps.

[1] *Facsimile Atlas*, p. 27, No. 33.

Ptolemy's Geography

Orto's maps had in the meantime been used in an Italian translation, by Leonardo Cernoti, of Magini's work, issued in folio by the Brothers Galignani at Venice in 1597-8. A second edition of Cernoti's work, in which the same sixty-four maps were used for the sixth time was published by the Galignanis at Venice in 1621, also in folio. Between these two Italian editions of Cernoti, there was the fourth Italian edition of Ruscelli of 1599 already mentioned. There were also the two large folio Latin editions issued at Amsterdam in 1605 and 1618-19, which require especial mention. The issue of 1605 was the first to exhibit the Greek and Latin text together, and it contained the twenty-eight maps of Mercator which had already been used in the Cologne edition of 1578 and 1584. In the preface, the publisher, Hondius, explains that the plates had been purchased from the heirs of Mercator. There are three copies of this fine edition in the Collection. One bears the imprint *Frankfurt* on the title, the second has *Amsterdam*, while the third has an additional printed title-page which appears to be hitherto unrecorded. In 1619 a very important edition was published by Hondius at Amsterdam in large folio, the work having been printed by Isaac Elzevir at Leyden. It contained both the Greek and Latin texts, and was edited by the celebrated geographer Petrus Bertius, who had also assisted in the preparation of the previous edition of 1605. The maps are extended to forty-seven in number, of which twenty-eight are those of Mercator which had previously been used in the editions of 1578, 1584 and 1605. Of the remainder, fourteen are by Abraham Ortelius, and had already done duty in various editions of his *Theatrum*. With this Amsterdam issue of

1619, and that of Venice of 1621, the long series of editions of Ptolemy's geography, revised and extended as text books of modern geography, seems to have come to an end. Afterwards we have only the issues 1695, 1698, 1704 and 1730 of the twenty-eight Mercator maps which applied solely to Ptolemy's ancient geography. The reason for the dropping of Ptolemy would seem to be that as the science of geography, and the knowledge of modern discoveries became more generally known and studied, modern textbooks came more and more into vogue. Thus on the ashes of a dead past rose the modern Atlases of Lafreri, Mercator, Ortelius, De Judaeis, Wytfliet, Blaeu, and many others, while Ptolemy became only a classic authority on ancient geography.

Of these modern Atlases only that of Wytfliet comes within the scope of this Collection, firstly, because he specially describes his work as a Supplement to Ptolemy, and secondly, because it is the first general Atlas of America. The popularity of the work must have been very considerable, for at least seven editions, in small folio, in Latin or French, appeared between 1597 and 1615, all of which are in this Collection. The first edition, which was in Latin, was printed at Louvain in 1597, and contained nineteen maps, all relating to America. The work was reprinted or reissued at Louvain in 1598, and again at Douay in 1603.

In 1605 it was translated into French and published at Douay with the addition of a second part relating to the East Indies, by Giovanni Antonio Magini of Padua and others. The nineteen maps of America were repeated, and four new ones of the East Indies added. A second French edition was issued at Douay in 1607, with the same twenty-three maps as in 1605, but with some additional text. The work was again issued in French at Douay in 1611, with the same twenty-three maps, but with the text somewhat differently arranged. In 1615 a new Latin edition was published at Arnhem by Joannes Jansson, with the original nineteen maps, but without the additional text and maps found in the French editions. This issue of 1615 appears to have been hitherto unrecorded. Both Winsor and Nordenskiöld[1] refer to an English edition of 1597, but so far this has escaped my observation.

Mention should here be made of a few other works which have been thought of sufficient collateral interest to be included in the Collection. Firstly, there is a fine copy in contemporary binding of Pomponius Mela's *De Situ Orbis*, printed at Vienna in 1518. This volume contains twenty-six manuscript maps of the same character as those illustrating Ptolemy's geography at this period. They were probably copied from an earlier Ptolemy MS., because, although somewhat similar, they do not agree entirely with the maps in any printed edition.

Secondly, we have the edition of Strabo, with Munster's Ptolemy maps as already referred to.[2]

Thirdly, the second edition of the famous English work on Navigation, Thomas Blunderville's *Exercises*, printed at London in 1597. This contains "a briefe description of universal Maps and Cards and of their use, and also the use of Ptolomey his Tables together with the true order of making the saide Tables," etc., etc.

[1] Winsor, *op. cit.*, p. 38; Nordenskiöld, *op. cit.*, p. 29.   [2] *Vide* p. 18.

Lastly, we have Abbé Halma's translation of part of Ptolemy's Greek text into French. The introduction and the appendices relating to the history and methods of ancient geography cause this book to be generally considered the best handbook to the scientific study of Ptolemy's Geography.

THE ULM PTOLEMY OF 1482

It now only remains to revert to a more particular description of the two copies of the Ulm edition of 1482 in the Collection, as already briefly referred to in their chronological sequence.[1] This edition exhibits many points of great interest to the bibliographer. Authorities agree in mentioning the fact that two issues of the book exist, but the details given are very meagre. In one, the recto of the last or colophon leaf is printed from a wood block, while in the other the same page is printed from type. The descriptions printed on the backs of the maps also exist in two distinct impressions, and the ornamental borders of these pages, which are of two distinct patterns, are often changed about or rearranged.

Several copies of the book, printed on vellum, are known, and it is believed that all these are exactly alike in the impressions of the map-descriptions, and also in the fact that the recto of the last leaf is printed from a woodblock. The copies on vellum also seem to agree exactly with those printed on paper, which have the woodblock last leaf; at least that is my experience after comparing seven

---

[1] *Vide* ante pp. 8-11.

copies. Judging from the sharper appearance of the ornamental borders to the map-descriptions in those copies with the woodcut end leaf, whether on vellum or paper, as compared with the borders in the issue with the type-printed end leaf, there can be little doubt that the former are the first issue.

But when we come to examine critically the issue with the end leaf printed from type, we immediately meet with a startling bibliographical and typographical puzzle which, it is believed, is here recorded for the first time. Out of twelve copies of the book personally examined, or of which written descriptions have been obtained, *no two have been found exactly alike*! That is to say, the above mentioned two separate impressions of the map-descriptions have been found mixed in various proportions in every copy of the book (with the type-printed end leaf) which has come under notice. Twenty-seven of these leaves seem to have been reprinted, the only one of which no reimpression has been found being Asia XII. Of the twelve copies of the book examined, one has as many as nineteen original impressions and eight reprints, while two others have only four originals and twenty-three reprints. The average of the twelve copies is eighteen or nineteen reprints out of the total of twenty-seven. Curiously enough the two copies containing twenty-three reprints do not agree, as each contains two not in the other copy. And amongst the twelve copies whenever the total number of reprints seems nearly to correspond, they are by no means the same leaves which constitute that total. How the sheets of the two impressions came to be mixed in such an extraordinary manner is a puzzle which has so far escaped a reasonable solution. In the twelve copies examined the total number found of each of the twenty-seven leaves was widely different, varying from four to eleven, the average being between eight and nine.

When endeavouring to tabulate the two states of the twenty-seven leaves, four copies were examined before I succeeded in securing particulars of them all. Excluding the two copies in the British Museum included in the four mentioned above, it would have taken the next seven copies examined to have made another perfect set. The whole twelve copies would only make four perfect sets if actually broken up for the purpose. I trust at some future time to publish tabulated descriptions of these curious and so far inexplicable variations.

The first copy in this Collection is of the earlier issue with the woodcut end leaf. The second copy has the type-printed end leaf, and sixteen original impressions of the map-descriptions, out of the twenty-seven. The remaining eleven leaves are reprints, as may be seen on comparison with the first copy, viz. Europe I, IV, VI, VII, New Italy, Africa II, III, Asia I (1st leaf), III, IV, and X. But this curious and interesting typographical puzzle by no means ends here, for in searching through the twelve copies to see if two could be found alike, a third distinct variation of three leaves was discovered. The first is merely a rearrangement of the border of Europe I. The second is a distinct re-impression of Africa II, first found in a copy purchased a few years ago for the John Carter Brown Library at Providence, and also found in the second copy in this Collection. The third is a similarly distinct reimpression of

Europe VI, so far only found in the second copy in this Collection.

The type used in the first and second issues is apparently the same throughout, but differently set up, but in these third impressions of Europe VI, and Africa II, the type of the capital letters is absolutely different, and may be described as of a fancy ornamental character in contradistinction to the usual plain Roman. The late Mr. Robert Proctor, the typographical expert at the British Museum, was greatly interested in this type. He informed me that it was entirely new to him, and after considerable search he failed to find it used in any other work. It was his intention to look fully into the matter after that fatal vacation of 1903, from which, alas, he never returned. He was inclined to think that this type might even represent an earlier impression than the other two varieties. This point I commend to typographical experts and bibliographers as well worthy of further investigation.

At all events, the borders of these two pages appear to be sharper than any appearing in either of the so-called first and second issues, and this fact points to an earlier impression.

Even now the most interesting and important new variation in regard to this remarkable book, remains to be announced and described.

In 1901, in the aforesaid copy purchased for the John Carter Brown Library, I discovered a map of the world which was entirely different, and evidently earlier than the usual one found in the work. In place of the usual straight lettering-scrolls along the tops or bottoms of the Heads of the Winds, the letterings are en-

Ptolemy's Geography

closed in scrolls partly encircling the heads in the form of halos. The usual engraver's inscription, "Insculptum est per Ioannẽ Schnitzer de Armszheim" does not appear, and the map does not contain so many names of places as the usual one. It is evidently from an entirely different and earlier woodblock. The importance of this map is readily to be perceived, for it is, as far as yet discovered, the earliest printed map in which an attempt is made to lay down the figure of the World on modern geographical conceptions beyond the limits of the old Ptolemeian maps. So far no other specimen of this valuable map has been found, but it is quite possible, now that it has been identified, others may be discovered hiding in copies of the work in out-of-the-way libraries. Very probably also, other leaves printed in the fancy type like Europe VI and Africa II may similarly be discovered now their existence has been pointed out.

I shall be extremely grateful to any person who will communicate to me particulars of any other copies of the early map or of additional leaves in the fancy type. It is my hope and intention, after further investigation, to publish facsimiles of the early map and fancy type leaves, together with a tabulated record of, and the necessary details for identifying, the variations of the map-descriptions as far as yet found. Some such table is of course indispensable for identifying the various states, as it is obviously impossible to get copies from various libraries together for comparison.

Mr. Eames,[1] quoting from Hoffmann's *Lexicon Bibliographicum*, records that a copy of this Ulm Ptolemy of 1482 on vellum is preserved in the Rhediger Library at Breslau, in which the name of the engraver is omitted from the map, and also the names of the place and printer from the colophon. I wrote to the librarian at Breslau to see if this

[1] *Op. cit.*

indicated another copy of the early map. In a most courteous reply he informed me that the Rhediger Library was now incorporated with the University, and that on a close examination of the map it was found to be merely the ordinary impression, but that the name of the engraver at the top had been cut off; further, that the omission in the imprint was presumably due to an erasure, traces of which were apparent.

In September, 1903, on the invitation of His Highness Prince Waldburg, I paid a visit to Wolfegg Castle, in company with my friends Mr. Soulsby, Superintendent of the Map Room in the British Museum, my partner Mr. Stiles, and the Rev. Prof. Joseph Fischer of Feldkirch, for the purpose of seeing the marvellous maps of Waldseemüller of 1507 and 1516, which the lucky hand of the learned Professor had recently awakened from the slumber of centuries in the old Castle.[1]

Amongst the numerous other treasures which the good and kindly Prince delighted in showing us, was a magnificent illuminated manuscript of Ptolemy's Geography on vellum, the maps of which the Professor had already identified as the prototypes of those in the Ulm printed volume of 1482. He gives a most interesting account of this MS. in his recent book *The Discoveries of the Norsemen*.[2]

If further proof were wanted that this MS. is the veritable prototype of the Ulm printed editions, it is found in the fact that the map of the world in the MS. agrees almost exactly with my early printed map now in the John Carter

---

[1] *Vide* the facsimiles as already cited on p. 12 *note*.
[2] *Norsemen*, p. 76 *note*, etc.

Brown Library. In both, the Heads of the Winds have the inscriptions encircling them instead of being in straight lines.

Prof. Fischer mentions that my printed map contains more names than in the Wolfegg MS., but the printed map has numerous insertions in contemporary manuscript which are almost indistinguishable from print in the small photograph which was then available. It would be interesting to actually compare the printed map with the Wolfegg Codex, but unfortunately the former had already gone to the John Carter Brown Library before my visit to Wolfegg.

Prof. Fischer gives a reduced facsimile of the Wolfegg MS. map in his book on the Norsemen, Plate V, which, until the facsimile of my printed map is published, will give a very good idea of its general characteristics, in contradistinction to the usual printed map found in the Ulm Ptolemys of 1842 or 1846.[1]

From the above brief notes it will readily be realized what an extremely important work the Ulm Ptolemy of 1482 really is, and what a field for further research it offers to the student, whether his interests be bibliographical, cartographical, or typographical.

HENRY N. STEVENS.

39, GREAT RUSSELL STREET,
LONDON, W.C.
20 *October*, 1908.

[1] *Vide* facsimile in Nordenskiöld's *Facsimile Atlas*, Plate XXIX.

EPILOGUE

In conclusion it is very gratifying to learn from Mr. Ayer, while this Essay has been in the Press, that when in the fullness of time the Collection must necessarily pass into other hands, he has already arranged for it to be preserved intact, by transferring it to the Newberry Library at Chicago. But let us hope he may long be spared to enjoy the personal possession of this grand set of books, which
in the distant future, as the
STEVENS-AYER PTOLEMY COLLECTION,
may well form a lasting memorial to
HENRY STEVENS OF VERMONT
who conceived the idea, and to
EDWARD EVERETT AYER OF CHICAGO,
by whose generous patronage it has been
brought to a successful
completion.

PTOLEMY'S GEOGRAPHY
A CHRONOLOGICAL LIST
OF THE EDITIONS
REPRESENTED
IN THE
STEVENS-AYER COLLECTION
WITH BRIEF DESCRIPTIVE NOTES
REPRINTED
FROM THE BOOKPLATES

AEVROPA
ASIA
AFRICA

CL. PT.
2nd century

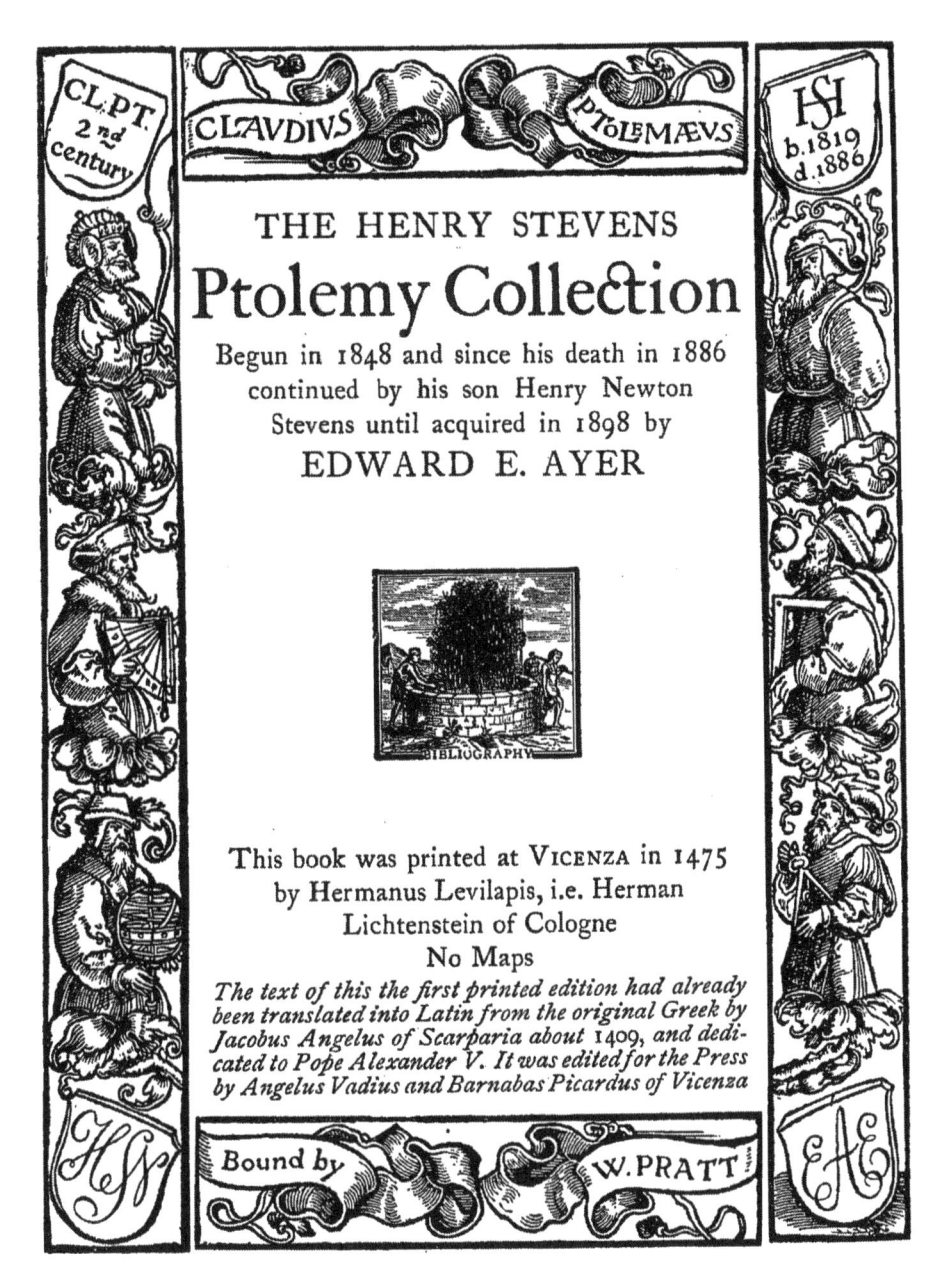

CLAVDIVS PTOLEMÆVS

HSt
b. 1819
d. 1886

THE HENRY STEVENS
Ptolemy Collection

Begun in 1848 and since his death in 1886 continued by his son Henry Newton Stevens until acquired in 1898 by

EDWARD E. AYER

BIBLIOGRAPHY

This book was printed at VICENZA in 1475 by Hermanus Levilapis, i.e. Herman Lichtenstein of Cologne
No Maps

The text of this the first printed edition had already been translated into Latin from the original Greek by Jacobus Angelus of Scarparia about 1409, and dedicated to Pope Alexander V. It was edited for the Press by Angelus Vadius and Barnabas Picardus of Vicenza

Bound by W. PRATT

A FACSIMILE OF THE FIRST BOOKPLATE

NOTA BENE

IN THE STEVENS-AYER COLLECTION EACH OF THE FOLLOWING NOTES APPEARED ON A SEPARATE BOOKPLATE IN THE STYLE OF THE FOREGOING SPECIMEN. THE TITLES AND COLLATIONS WERE NOT GIVEN FOR THE OBVIOUS REASON THAT THEY COULD BE SEEN IN THE BOOKS THEMSELVES. FOR FULL TITLES AND COLLATIONS, WITH FULLER DESCRIPTIONS THAN SPACE PERMITTED TO BE GIVEN ON THE BOOKPLATES, THE READER IS REFERRED TO THE BIBLIOGRAPHIES OF EAMES, NORDENSKIÖLD AND WINSOR ALREADY CITED

(*Vide Notes on pp.* 3, 10 *and* 14 *ante*).

A CHRONOLOGICAL LIST OF THE PRINTED EDITIONS OF PTOLEMY'S GEOGRAPHY
1475—1730

ALL IN LATIN UNLESS OTHERWISE DESCRIBED

[1475 *folio*]
Printed at VICENZA in 1475 by Hermanus Levilapis, *i.e.*, Herman Lichtenstein of Cologne
No Maps

THE text of this, the first printed edition, had already been translated into Latin from the original Greek by Jacobus Angelus of Scarparia, about 1409, and dedicated to Pope Alexander V. It was edited for the Press by Angelus Vadius and Bernardus Picardus of Vicenza.

[1478 folio]
Printed at ROME in 1478 by Arnoldus Buckinck
27 Copperplate Maps

THE translation of Jacobus Angelus, edited, with the emendations of Georgius Gemistus, by Domitius Calderinus of Verona. The first edition with maps. An account of the origin and progress of this edition is given in the Preface. The maps are generally supposed to be the first examples of copperplate engraving for books. This volume is the only known work bearing the imprint of Buckinck.

[c. 1480 folio]
Believed to have been printed at FIRENZE (Florence) by Nicolo Todescho about 1480
31 Copperplate Maps on 60 leaves
[In Italian]

THE excessively rare first issue of Francesco Berlinghieri's metrical paraphrase of Ptolemy's Geography, having the recto of the first leaf and the entire last leaf blank. A later issue has a Register with an undated colophon of Nicolo Todescho printed on the recto of the last leaf, blank in this first edition; whilst a considerably later re-issue has the further addition of a new title-page in red ink printed on the recto of the front leaf previously blank. A copy of this third issue is also in the Collection. Some authorities think the first edition may have been published as early as 1477, in which case the maps may be earlier than those of the Rome edition of 1478, from which they are quite distinct in character and inferior in the quality of the engraving.

❦

[*c.* 1480 (re-issued *c.* 1500) *folio*]
Printed at FIRENZE (Florence) by Nicolo Todescho about 1480, and re-issued, probably after 1500, with a title-page
31 *Copperplate Maps on* 60 *leaves*
[*In Italian*]

A RE-ISSUE of Francesco Berlinghieri's metrical paraphrase of Ptolemy's Geography first published about 1480, a copy of which is in the Collection. In the original edition the recto of the first leaf and the entire last leaf were blank. This re-issue contains a title-page in red ink printed on the recto of the original first leaf, and a Register and Todescho's Colophon on the recto of the last leaf. According to the Register the first leaf should be blank (" Prima Alba"), from which it would appear that at an intermediate period copies were issued with the Register before the title-page was finally added. Todescho is believed to have printed between 1477 and 1486, so that the Register was presumably added before the latter date. But the title-page was probably not added till after 1500, as the type is unquestionably more modern in character than that of the rest of the book. The text and the Maps are the same as the original issue.

40   *A Chronological List of the Printed*

[1482 (first issue?) *folio*]

Printed at ULM in 1482 by Leonardus Hol

32 *Woodcut Maps*

THE translation of Jacobus Angelus, edited by Donnus Nicolaus Germanus, who also redrew, corrected and improved the Maps and added 5 new modern ones, of which one contains the first printed representation of Greenland. This book exhibits many points of interest to the bibliographer. The recto of the last leaf is in some copies (like this) printed from a wood-block, while in others (like the second copy in the Collection) it is printed from type. All the descriptions printed on the backs of the maps exist in two (and in some cases three) distinct impressions. In all copies with the woodcut end leaf the map descriptions are of the same impression, presumably the first; but in those copies with the printed end leaf, the impressions are unaccountably mixed in various proportions as more particularly noted on the Book-Plate to the other copy in this Collection.

[1482 (second issue?) *folio*]

Printed at ULM in 1482 by Leonardus Hol

32 *Woodcut Maps*

THE translation of Jacobus Angelus edited by Donnus Nicolaus Germanus, who also redrew, corrected, and improved the Maps, and added 5 new modern ones, of which one contains the first printed representation of Greenland. In this issue the recto of last leaf is printed from type, whereas

Editions of Ptolemy's Geography

in the first copy in the Collection it is printed from a woodblock. All the descriptions on the backs of the Maps exist in two (and in some cases three) distinct impressions. In copies with the woodcut end leaf the descriptions appear to be of the first issue and all copies are believed to be alike; but in this second issue the various impressions are unaccountably mixed in various proportions. In this copy eleven leaves will be found to differ from the first issue. The leaves Europe VI and Africa II are of a third variety, and so far no other copy of Europe VI has been observed and only one other copy of Africa II.

[BOLOGNA 1462 (1482?) *folio*]

[N.B.—This edition not being represented in the Collection, the reader is referred to the description given by Mr. Eames (*vide ante*, p. 3, *note*).]

[1486 *folio*]
Printed at ULM in 1486 for Justus de Albano de Venetiis by Johann Reger
32 *Woodcut Maps*

THE translation of Jacobus Angelus, edited by Nicolaus Donis. The second edition of Donis, with the addition of an Index in 42 leaves and a new treatise "De Locis" in 24 leaves. The maps are from the same blocks as the first Donis edition of 1482.

[N.B.—Since this Note was prepared for the Bookplate, Prof. Fischer has shown that "Donis" has for centuries been an erroneous citation of *Donnus* Nicolaus Germanus. (*Cf.* p. 11, *ante*, and Fischer *Norsemen*, p. 72, etc.)]

[1490 *folio*]
Printed at ROME in 1490 by Petrus de Turre
27 *Copperplate Maps*

THE translation of Jacobus Angelus, accompanied by the Index and Treatise of Nicolaus Donis (see Ulm, 1486). The second Rome edition. The maps are from the same plates as those in the edition of 1478.

[N.B.—As to "Donis," see Note to 1486.]

[1507 *folio*]
Printed at ROME in 1507
by Bernardinus Venetus de Vitalibus
33 *Copperplate Maps (and the Ruysch Map)*

A NEW edition of the translation of Jacobus Angelus, revised and edited by Marcus Beneventanus and Joannes Cota of Verona. The 27 maps from the editions of 1478 and 1490 are repeated and 6 new modern maps added. This copy has the advantage of being one of those occasionally found, in which is inserted an impression of the famous Ruysch map, which is believed not to have been really issued till 1508. (See 1508.) It is the second map in this book.

[1508 *folio*]
Printed at ROME in 1508
by Bernardinus Venetus de Vitalibus
34 *Copperplate Maps*

A RE-ISSUE of the edition of 1507, with a new title-page and the addition of Beneventanus's description of the New World in 14 leaves, and the new Map of the World by Johann Ruysch. This is the first edition containing any account of the New World, and Ruysch's map is (as far as has yet been definitely ascertained) the first printed map delineating any part of the New World. It is the second map in the book.

[N.B.—Since this Note was prepared for the Bookplate, the Waldseemüller Map of 1507 has been discovered (*vide ante*, p. 12, note). Ruysch's Map can, therefore, now only be described as the first in any edition of Ptolemy's Geography to delineate any part of the New World.]

[1511 *folio*]
Printed at VENICE in 1511
by Jacobus Pentius de Leucho
28 *Woodcut Maps*

THE version of Jacobus Angelus edited and remodelled by Bernardus Sylvanus of Eboli. The maps are entirely fresh wood-blocks, with the letterings type-printed in red and black. The edition is principally esteemed for the New

World Map (the last in the book), drawn on a heart-shaped projection, which contains the first printed delineation of any part of the North American continent.

[N.B.—For reasons given in footnote to 1508, the words "in any edition of Ptolemy's Geography" must now be added to the above.]

[1512 4*to*]

[Stobnicza's Introduction to "Ptolemy's Geography," First Edition printed at Cracow in 1512. No copy in the Collection. *Vide* 1519 for second edition. Cf. *ante*, pp. 13-14.]

[1513 *folio*]

Printed at STRASSBURG in 1513 by Joannes Schott

47 *Woodcut Maps*

PERHAPS the most important of all the editions. It was commenced by the famous St. Dié geographer, Martin Waldseemüller, possibly as early as 1505, and, after much delay, finally edited and published by Jacobus Eszler and Georgius Ubelin of Strassburg in 1513. The maps are all from fresh wood-blocks, and 20 new modern maps in a Supplement here first appear. These include a modern Map of the World, "Orbis Typus universalis," and the "Tabula Terre Nove," the latter being one of the earliest maps devoted specially to the New World.

[1514 *folio*]
Printed at NUREMBERG in 1514 by Joannes Stuchs
No Maps

A NEW translation by Joannes Werner, with paraphrases and annotations, and with the addition of other geographical treatises by Werner, Georgius Amirutzes and Regiomontanus.

[1518 *folio*]
Printed at VIENNA in 1518 by Joannes Singrenius for the famous publisher Lucas Alantsius
26 *Maps added in Manuscript*

A FINE copy in contemporary binding of the "Libri de situ orbis tres" of Pomponius Mela, edited with annotations by Joachim Vadianus. This volume is included in the Ptolemy Collection because it contains at the end 26 maps in manuscript of the same character as those illustrating Ptolemy's Geography at this period. They have probably been copied from an earlier Ptolemy MS., because, although very similar, they do not agree entirely with any printed edition. They comprise the usual 10 maps of Europe, 4 of Africa, and 12 of Asia, but the map of the World is not included.

[1519 4*to*]
Printed at CRACOW in 1519 by Hieronymus Vietor
Woodcut Map in two sheets (in facsimile)

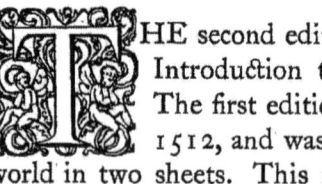

THE second edition of Joannes de Stobnicza's Introduction to Ptolemy's Geography, etc. The first edition was printed at Cracow in 1512, and was accompanied by a map of the world in two sheets. This map has always created great interest because, until the discovery in 1901 of the original Waldseemüller map of 1507, it was the earliest known in which the North and South American continents were joined. The Stobnicza map turns out to be a slavish copy of Waldseemüller's (cf. the Waldseemüller map facsimiles). It is doubtful whether the map was reissued with this second edition of Stobnicza's text, as no known copy contains it; but the facsimiles here added will enhance the interest of this volume.

[Cf. *ante*, pp. 13-14.]

[1520 *folio*]
Printed at STRASSBURG in 1520 by Joannes Schott
47 *Woodcut Maps*

SECOND Strassburg edition, edited by Georgius Ubelin, one of the editors of the 1513 edition, of which this is mainly a reprint with certain omissions.

[1522 *folio*]
Printed at STRASSBURG in 1522 by Joannes Grüninger
50 *Woodcut Maps*

THE third Strassburg edition, edited by Laurentius Fries or Phrisius. The maps are re-engraved on a smaller scale, and three new ones added. The last is remarkable as the first map bearing the name "America" that appeared in Ptolemy's work.

[1525 *folio*]
Printed at STRASSBURG in 1525 by Joannes Grüninger
50 *Woodcut Maps*

THE fourth Strassburg edition, re-translated by Bilibald Pirckeymher, with the annotations of Regiomontanus, and said to have been edited by Johann Huttich. The maps are the same as those in the 1522 edition.

[1533 4*to*]
Printed at BASLE in 1533 by Hieronymus Froben
No Maps

THE first edition of the Greek Text of Ptolemy's "Geography."

[1535 *folio*]
Printed at Lyons in 1535
by Melchior and Gaspar Trechsel·

50 Woodcut Maps

EDITED by Michael Villanovanus, better known as Servetus. The woodcut borders and ornaments are said to be the work of Hans Holbein and Graf of Basle. Many copies of the book are said to have been burned by the order of Calvin at the time of the execution of its editor Servetus in 1553. The maps are the same as in the editions of 1522 and 1525.

[1540 *folio*]
Printed at Basle in 1540 by Henricus Petri

48 Woodcut Maps

NEW and important edition, revised and edited by Sebastian Munster, who designed the maps anew and added a geographical appendix.

[1540 *sm. 8vo*]
Printed at Cologne in 1540
by Joannes Ruremundanus

No Maps

NEW Latin translation from the Greek text, by Joannes Noviomagus (Johann Bronchorst).

Editions of Ptolemy's Geography

[1541 folio]
Printed at VIENNE in Dauphiné in 1541 by Gaspar Trechsel

50 Woodcut Maps

THE second edition, edited by Villanovanus (Servetus), newly revised and corrected. The maps are from the same blocks as the editions of 1522, 1525 and 1535.

[1542 folio]
Printed at BASLE in 1542 by Henricus Petri

48 Woodcut Maps

THE second edition of Sebastian Münster, reprinted from the edition of 1540 without alteration or addition. The maps are from the same blocks, but the descriptions have been reset and the borders changed about.

[1545 folio]
Printed at BASLE in 1545 by Henricus Petri

54 Woodcut Maps

THE third edition of Sebastian Münster, with 6 new maps added. The remaining 48 maps are from the same blocks as the editions of 1540 and 1542, but with the descriptive text reset, and the borders changed about.

[1548 sm. 8vo]
Printed at VENICE in 1548 by Nicolo Bascarini
60 Copperplate Maps

THE first edition in Italian. Translated by Pietro Andrea Mattioli of Siena. The maps designed by Jacopo Gastaldo, mostly after those of Münster, but with many important additions.

[1552 folio]
Printed at BASLE in 1552 by Henricus Petri
54 Woodcut Maps

THE fourth edition of Sebastian Münster, with an additional treatise and enlarged indexes. The maps are from the same blocks as the edition of 1545 with some of the letterings and descriptive text altered.

[1561 4to]
Printed at VENICE in 1561 by Vincenzo Valgrisi
64 Copperplate Maps

A NEW and important edition in Italian, with a new series of maps. Edited with a long "Exposition" by Girolamo Ruscelli, and with a further addition of a "Discourse" by Gioseppe Moleto.

[1562 4to]
Printed at VENICE in 1562 by Vincenzo Valgrisi
64 Copperplate Maps

A NEW edition, revised and annotated by Josephus Moletius. The maps are the same as in the Italian edition of 1561.

[1564]
Published at VENICE in 1564 by Giordano Ziletti
64 Copperplate Maps

IN Italian. The second edition of Girolamo Ruscelli, the first having been issued in 1561 by Vincenzo Valgrisi. Ziletti also published in this same year (1564) the second edition of the Latin Ptolemy edited by Josephus Moletius. Both the Latin and Italian editions of this year are very scarce.

[1564 4to]
Published at VENICE in 1564 by Giordano Ziletti
64 Copperplate Maps

THIS edition is so rare that neither Eames nor Winsor mention it in their Bibliographies. It is a re-issue, by a fresh publisher, of the new Latin edition of Josephus Moletius first published at Venice by Vincenzo Valgrisi in 1562.

Only the first sheet, comprising the Title and Dedication, was reprinted; the editor's name being omitted from the title-page. Ziletti also published in this same year the second edition of Ruscelli's Ptolemy in Italian, which is also extremely rare.

[1571 *folio*]

Printed at BASLE in 1571 by Henricus Petri

24 Woodcut Maps (+ 3 *repeated* = 27)

THIS Latin edition of Strabo's "Geography" is included in this Collection because it contains 24 of the Maps designed by Sebastian Münster to illustrate Ptolemy's "Geography," and which were used in his four editions of 1540, 1541, 1545 and 1552, all by the same printer, Petri. Münster died in 1552, and after nearly twenty years it is curious to find the same maps from the same wood-blocks again appearing in an entirely different work by the same printer. The famous Greek geographer, Strabo, died about A.D. 20.

[1574 4*to*]

Printed at VENICE in 1574 by Giordano Ziletti

65 Copperplate Maps

IN Italian. The third edition of Girolamo Ruscelli, revised and corrected by Gio. Malombra. One new map is added (Roma); the others are the same as in the editions of 1561, 1562 and 1564.

[1578 *folio*]
Printed at COLOGNE in 1578 by Godefridus Kempen
28 *Copperplate Maps*

THE first edition of Mercator's Maps for Ptolemy's "Geography," but issued without the text.

[1584 *folio*]
Printed at COLOGNE in 1584 by Godefridus Kempen
28 *Copperplate Maps*

THE first edition of Ptolemy's "Geography" with Mercator's Maps. The text is the version of Pirckeymher, edited by Arnoldus Mylius. The maps are the same as in the edition of 1578.

[1596 *4to*]
Printed at VENICE in 1596
by the Heirs of Simone Galignani de Karera
64 *Copperplate Maps*

A NEW (Latin) edition with new maps, edited by Giovanni Antonio Magini of Padua.

A Chronological List of the Printed

[1597 4*to*, first copy]
Printed at COLOGNE in 1597 by Petrus Keschedt
64 *Copperplate Maps*

THE second edition of Giovanni Antonio Magini of Padua. The maps are the same as in his first edition of 1596.

[1597 4*to*, second copy]
Printed at COLOGNE in 1597 by Petrus Keschedt
64 *Copperplate Maps*

THE second edition of Giovanni Antonio Magini of Padua. The maps are the same as in his first edition of 1596. This copy has the additional imprint on the verso of the last leaf: "Arnhemii, Apud Joannem Jansonium Bibliopolam Anno MDXCVII." The other copy in the Collection does not bear this imprint.

[1597 *folio*]
Printed at LOUVAIN in 1597 by Joannes Bogardus
19 *Copperplate Maps all relating to America*

THE first edition in Latin of Cornely Wytfliet's Supplement to Ptolemy's "Geography." It relates entirely to America, and is the first general American Atlas.

Editions of Ptolemy's Geography

[1597 *4to*]
Printed at LONDON in 1597 by John Windet
Numerous woodcut diagrams (some movable)

THE second edition of Thomas Blundeville's "Exercises," the best English work of the period on Astronomy, Geography and Navigation. Seven editions appeared between 1594 and 1636. The book finds a place in the Ptolemy Collection by reason of the contents of folios 362-392: "A briefe defcription of univerfall Maps and Cards and of their ufe, and alfo the ufe of Ptolomey his Tables together with the true order of making the faide Tables, and of all other Mappes and Cardes as well univerfall as particular."

[1597-8 *folio*]
Printed at VENICE in 1597-8
by the Brothers Galignani
64 *Copperplate Maps*

A TRANSLATION into Italian by Leonardo Cernoti from Magini's Latin edition of 1596, and with the same maps.

[1598 *folio*]
Printed at LOUVAIN in 1598 by Gerardus Rivius
19 *Copperplate Maps the same as in the edition of* 1597

THE second edition in Latin of Wytfliet's Supplement to Ptolemy's "Geography."

A Chronological List of the Printed

[1598-9 4*to*]

Printed at VENICE in 1598-9
by the Heirs of Melchior Sessa

69 *Copperplate Maps*

IN Italian. The fourth edition of Ruscelli's Ptolemy, revised, enlarged, and edited by Gioseppe Rosaccio. The maps are mostly from the same plates as the editions of 1561, 1562, 1564 and 1574, but the plates have been retouched and figures of ships, fishes, etc., introduced. Five entirely new maps have also been added.

[1603 *folio*]

Printed at DOUAY in 1603 and published
by François Fabri

19 *Copperplate Maps as in the editions of* 1597 *and* 1598

ALTHOUGH called Second Edition on the title, this appears really to be the third edition of Wytfliet's Supplement to Ptolemy's " Geography."

[1605, first copy, *folio*]

Printed in 1605 at the expense of Cornelius Nicolaus and Jodocus Hondius, and published at AMSTERDAM and FRANKFURT. Copies occur with the imprint of either town. This has Frankfurt

28 Copperplate Maps

THE first edition of the Greek and Latin text together. Edited by Petrus Montanus. The Preface by Hondius contains an account of the origin of this edition. The maps are the same as in the editions of 1578 and 1584.

[1605, second copy, *folio*]

Printed in 1605 at the expense of Cornelius Nicolaus and Jodocus Hondius, and published at AMSTERDAM and FRANKFURT. Copies occur with the imprint of either town. This has Amsterdam, but a copy of the Frankfurt issue is also in the Collection, as well as a third copy containing a printed title-page in addition to the engraved one

28 Copperplate Maps

THE first edition of the Greek and Latin text together. Edited by Petrus Montanus. The Preface by Hondius contains an account of the origin of this edition. The maps are the same as in the editions of 1578 and 1584.

[1605, third copy, *folio*]

Printed in 1605 at the expense of Cornelius Nicolaus and Jodocus Hondius, and published both at AMSTERDAM and FRANKFURT. Copies with the imprint of each town are in the Collection. This copy bears the name of both places on the engraved title, and has, in addition, a printed title-page with the imprint of Amsterdam, 1605. This printed title appears to be unknown to bibliographers. Judging by the type it appears to be of a somewhat later date, and may possibly have been used for a re-issue of the original stock

28 *Copperplate Maps*

THE first edition of the Greek and Latin text together. Edited by Petrus Montanus. The Preface by Hondius contains an account of the origin of this edition. The Maps, by Mercator, are the same as in the editions of 1578 and 1584.

[1605 *folio*]

Printed at DOUAY in 1605 and published by François Fabri

19 *large Copperplate Maps in the First Part and 4 small ones in the Second Part*

A TRANSLATION of Wytfliet's Supplement to Ptolemy into French, with the addition of a second part by Giovanni Antonio Magini of Padua and others, relating to the East Indies, with 4 new maps. Scarce edition, not mentioned by Winsor.

[1607 *folio*]
Printed at DOUAY in 1607 and published
by François Fabri
19 *large Copperplate Maps in the First Part and
4 small ones in the Second Part*

WYTFLIET'S Supplement to Ptolemy and Magini's "History of the East Indies" translated into French and much enlarged from the edition of 1605, with a new treatise, entitled "La Suite de l'histoire des Indes Orientales."

[1608 *4to*]
Printed at COLOGNE in 1608 by Petrus Keschedt
64 *Copperplate Maps*

THE third edition of Giovanni Antonio Magini of Padua. The maps are the same as in his first edition of 1596 and second of 1597.

[1611 *folio*]
Printed at DOUAY in 1611 and published
by François Fabri
19 *large Copperplate Maps in the First Part and
4 small ones in the Second Part*

WYTFLIET'S Supplement to Ptolemy and Magini's "History of the East Indies" translated into French. In this edition the "Discours de la Conversion des Indiens occidentaux" is placed at the end of the first part, instead of at the end of the third part as in the edition of 1607.

A Chronological List of the Printed

[1615 *folio*]
Printed at ARNHEM in 1615 by Joannes Jansson
19 *Copperplate Maps as in the editions
of* 1597, 1598 *and* 1603

A SCARCE edition of Wytfliet's Supplement to Ptolemy, not mentioned by Winsor. According to the title it purports to be newly revised and corrected.

[1617 4*to*]
Published at ARNHEM in 1617 by Joannes Jansson
64 *Copperplate Maps*

THE fourth edition of Giovanni Antonio Magini of Padua, edited by Gaspar Ens. The maps are the same as in the preceding editions of 1596, 1597, and 1608.

[1618-19 *folio*]
Printed at LEYDEN in 1618-19 by Isaac Elzevir and published at AMSTERDAM by Jodocus Hondius
47 *Copperplate Maps*

AN important edition, re-edited by Petrus Bertius, with many new maps and additions. Contains the Greek and Latin text in parallel columns. 28 of the maps are Mercator's, the same as in the editions of 1578, 1584, and 1605, while 14 of the new maps are by Ortelius, from various editions of his "Theatrum."

[1621 *folio*]
Printed at PADUA in 1621 by the Brothers Galignani
64 *Copperplate Maps*

THE second edition of Cernoti's translation into Italian of Magini's Ptolemy. The maps are the same as in the first edition of 1597-8. The present copy is remarkable as being perfectly uncut.

[1695 *folio*]
Published at FRANEKER and UTRECHT in 1695
28 *Copperplate Maps*

A NEW issue of the 28 Mercator maps without any text. Printed apparently from the old plates, but with the titles re-engraved, and with some alterations in the ornamentation. The border to the World map is entirely re-engraved.

[1698 *folio*]
Published at UTRECHT and FRANEKER in 1698
28 *Copperplate Maps*

APPARENTLY a re-issue of the edition of 1695 with a new imprint.

[1704 *folio*]

Published at AMSTERDAM and UTRECHT in 1704

28 *Copperplate Maps*

APPARENTLY a re-issue of the editions of 1695 and 1698 with a new imprint. Some copies have LUGDUNI BATAVORUM, etc., on a slip pasted over the Amsterdam imprint.

[1730 *folio*]

Published at AMSTERDAM in 1730

28 *Copperplate Maps*

A RE-ISSUE of the 28 Mercator Maps as they appeared in the editions of 1695, 1698 and 1704, with the addition of a copious index. This is the last edition of Ptolemy till quite recent times.

[1828]

Printed at PARIS in 1828

CONTAINS the Greek Text of Ptolemy's Geography translated into French by Abbé Halma. The introduction and the appendices relating to the history and methods of ancient geography cause this work to be generally considered the best handbook to the scientific study of Ptolemy's Geography.

CHISWICK PRESS: CHARLES WHITTINGHAM AND CO.
TOOKS COURT, CHANCERY LANE, LONDON.